Siegfried Genreith

The Source of the Universe

Intelligent decision making, quantum measurements and gravity are three different traits of a single flame-like random process. Is consciousness the true foundation of the universe?

General relativity and quantum mechanics describe seemingly incompatible traits of our universe. Their unification into a theory-of-everything challenged physics for the last century.

The Genl model satisfies both fundamental theories. It comprises a random walk process operating on a swarm-like construct and implements the competition among a finite set of ideas. Genl precisely fulfils the predictions of quantum measurements while its dynamics locally satisfy Einstein's field equation. The model suggests, that the perceivable universe is evolving according to the collapse of its quantum state rather than a smoothly evolving wave function as widely believed in modern physics. Consequently, gravitation cannot be directly derived from quantum mechanics or vice versa. Both simply describe distinct perspectives onto the previously unknown swarm-like decision process operating at the very basis of our universe.

Genl requires a fundamental change of perspective. In particular, within the model gravity is only a side effect of an evolutionary selection pressure.

Siegfried Genreith studied mathematics at the University of Cologne. During his professional career at a world-renowned IT company, he developed applications in the field of artificial intelligence in the late 1980s. The study of the scientific fundamentals of intelligence and consciousness has for many years become his passionate determination. Siegfried has three adult children and lives in the beautiful landscape of the German Eifel.

© 2017 Siegfried Genreith

Herstellung und Verlag: BoD - Books on Demand, Norderstedt

ISBN 9 783848 223572

Bibliografische Information der Deutschen Nationalbibliothek: Die Deutsche Nationalbibliothek verzeichnet diese Publikation in der Deutschen Nationalbibliografie; detaillierte bibliografische Daten sind im Internet über dnb.dnb.de abrufbar.

Contents

1 **Introduction**   3

2 **Results**   5
    2.1   The Genl model 5
    2.2   QM measurement statistics 8
    2.3   Spacetime metric 9

3 **Discussion**   11

4 **Methods**   15
    4.1   P-swarms 15
    4.2   The Genl process on two options 16
    4.3   E-swarms 18
    4.4   The Genl process on E-swarms 20
    4.5   Transformations 21
    4.6   Restarting the process 22
    4.7   Process statistics 23
    4.8   A spacetime geometry 25
    4.9   Data availability 30

5 **Tables**   31

6 **Figures**   33

1 Introduction

Finding a theory of everything (ToE) that harmonizes the stochastic, microscopic quantum world with the deterministic, perceivable universe poses one of the major challenges in modern physics. Numerous ToE candidates have been brought forward in the past, most popular the diverse flavours of supergravity-, string- and M-theory [1, 2, 3]. However, all these theories are based on purely speculative assumptions (e.g. the 11 dimensions proposed in M-theory), have questionable predictive power and limited theoretical ground. These ToE candidates further suffer from their immense mathematical complexity and the requirement for finest parameter tuning in order to mimic only selected aspects of quantum mechanics (QM) and Einstein's general relativity theory (GRT). Unsurprisingly, none of the existing ToE[4] candidates is even close to being experimentally verifiable with physical methods of today or the foreseeable future. Expanding and refining existing ToE proposals is unlikely to overcome these profound obstacles, but rather prone to ever increasing mathematical complexity and therefore room for pure speculation.

In principle, a ToE should not only explain gravity and quantum mechanics out of one single model, but every physical phenomenon in the universe. Several authors have suspected a fundamental relationship among at first sight very different domains: Intelligent decision making [5, 6, 7, 8, 9], the mysterious quantum measurement process and gravitation in terms of Einstein's general relativity theory (GRT)[10, 11]. Consequently, all three of the aforementioned domains might represent a valid entry point to some kind of theory of everything (ToE) [12, 13]. However, so far, the problem of emergent reality[14, 15] has been tackled almost exclusively from the quantum perspective, while the other two, possible routes were largely ignored (Figure 1).

In this work, I present GenI ([dʒiːnaɪ] for generic intelligence), a swarm-like stochastic model inspired by artificial intelligence (AI) that transcends the gap between GRT and QM and unlocks a completely new path towards a ToE. GenI operates on a swarm-like construct and implements a competition of ideas resulting in a rather chaotic selection process directed by a small set of simple rules. Remarkably, the probability distribution for idea selection only depends on the system's state at the beginning and precisely matches quantum measurement predictions. At the same time, the dynamics of the GenI process locally follow geodetic lines in a four dimensional Riemann space whose metric fully satisfies the requirements of Einstein's field equation. Besides the demonstrated theoretical validity, its mathematical simplicity, robustness and possible experimental accessibility distinguish GenI from all established ToE candidates.

The GenI model proposes a simple and elegant solution to the GRT vs QM dilemma and thereby unveils a fundamentally different perspective onto our universe.

2 Results

2.1 The Genl model

Biological swarms are comprised of relatively simple individuals that collectively perform surprisingly well in solving complex problems, e.g. maximizing survival and reproduction.

Fish swarms, for instance, decide on their movement by following a simple set of local rules generally obeyed by each swarm member [16]. From time to time, however, individuals make mistakes and disrespect the swarm's inner rule set. Remarkably, this seemingly imperfection appears to be a key ingredient of a swarms success in using and exploring its environment. With a certain probability, the swarm will follow its runaway individuals, resulting in sudden, highly dynamic changes in the swarm shape and swimming direction. In case the unexpected turn appears to improve the swarm's well-being or chance of survival it can keep going and explore. Hence, random mistakes by individuals enable the swarm to transcend the boundaries imposed by its own, inner laws, providing an enormous flexibility and exploratory potential eventually enabling creative problem solving.

Genl mimicks this swarm-like decision making strategy. In strong contrast to previous AI algorithms inspired by collective swarm intelligence, however, Genl implements a by far higher, so-far unexplored degree of randomization and uses an innovative swarm formalization introduced below.

The exclusive purpose of a Genl swarm S is to make a decision from a given set of options. The Genl swarm consists of an arbitrary but finite number of abstract individuals s_j. "Abstract" here means that swarm members are defined by a set of type-specific characteristics. You may imagine each member e.g. as comprising a infinitesimally small sphere with an individual number on it.

Alike biological swarms, the Genl swarm reacts to changes in its environment. The way the Genl swarm responds is thereby defined by the Genl process.

The size of the swarm as well as its individuals will change dramatically over time. Thus, a swarm behaves somewhat similar to a flame, constantly sucking "material" from its surroundings and "burning" it in order to grow or shrink, while dynamically changing its shape.

In the following, I will repeatedly return to this picture of a flame to illustrate GenI's basic features.

The shape of a swarm at a specific moment is referred to as "state". The state hides most of the swarm's inner swarm complexities and is represented by an array of n complex numbers, which I will refer to as the "amplitudes". "n" thereby determines the maximum number of possible choices the swarm may decide about. It is calculated by mapping each swarm member to a complex vector which are then all summed up.

Importantly, a swarm incorporates a significant number of entities that are hidden from the state view, as for any pair $s, t \in S$ with $s \mapsto w$, $t \mapsto -w$ the aforementioned sum is zero. A maximal set N_S of such *null pairs* represents the *entropy* of the swarm and $S \backslash N_S$ is a *entropy freed swarm*. For large swarms these null pairs are typically in the majority (Figure 2). They serve as "burning material" to keep the "flame" alive. Importantly, a real flame also needs continuous oxygen supply to keep the burning reaction going.

The corresponding counterpart ingredient in GenI is the *excitation*. Excitation, which is provided by the *environment* defined by a set of n independent vectors $B_E = \{a_j : j = 1 \ldots n\} \subset \mathbb{C}^n$. These vectors represent the possible options a_j the swarm S can choose from. The resulting swarm's shape is dependent on the environment and defined by $state(S) = \sum_{j=1}^{n} \beta_j a_j$. The component $\beta_j a_j$ refers to the *idea* of S about option j and $\beta_j \in \mathbb{C}$ its *amplitude*. Taken together, the state of a swarm represents a superposition of ideas about a given set of options defined by the environment (see also [17]).

The entropy is essential to keep the GenI process running with high volatility. It is responsible for the seemingly chaotic characteristics of the GenI decision making process.

GenI swarms can generally be classified into *P-swarms* (for Pauli-swarm) and *E-swarms* (for eigenvector-swarm).

The P-swarms implement decision making based on only two possible options (YES-NO type). This is the basic Genl swarm type that demonstrates the full power of the Genl model. Its members each map to one of the 16 elements of the well known Pauli group denoted by $\{i^k p_j : j, k = 0 \ldots 3\}$. Thus the swarm individuals are grouped into a few image types p_j, where p_0 represents the identity in $Mat(2x2, \mathbb{C})$, i the imaginary unit and p_1, p_2, p_3 the three Pauli matrices.

Any P-swarm S thus receives an image in a two dimensional complex matrix algebra. A linear *perspective* transformation maps the resulting operator onto a two dimensional complex state. This is achieved by right multiplying a fixed perspective vector $v \in \mathbb{C}^2$. So the two dimensional state of the swarm is determined by Sv. The operator image of S on the other hand can be used to derive an internal environment in terms of its inbound symmetries. Such an environment is sometimes referred to as an *observable*, which is the usual notation in QM for a hermitian operator. The observables' (orthonormal) eigenvector base then defines the environment introduced above.

E-swarms are leveraged to implement decision making among many possible options. Its individuals map directly to vectors. A perspective transformation is not needed here and no operator image gets exposed to determine any internal environment. In case of two choices E- and P-type swarms are typically equivalent with respect to the statistically expected outcome of the Genl process.

The dynamical features of the model are governed by the Genl process. It operates both at the swarm level and at the state level and implements a competition amongst ideas of a swarm
$S = \sum_{j=1}^{n} \beta_j a_j$. During the decision making process, the swarm carries out a random walk within a given environment constrained by only three simple, local rules (Figure 3): 1. Increase entropy, 2. lower the excitation value, and 3. disregard the rules whenever you want (see below).

Importantly, the fail rate of meeting the second rule is high, enabling the swarm to carry out unpredictable walks and fully explore the environment (Figure 4).

A full iteration of the Genl algorithm starts by adding entropy to each idea according to its amplitude β_j. Next, the corresponding excitation values are calculated as $\epsilon_j(S)^2 := 4\frac{|\beta_j|^2 \sum_{k\neq j}|\beta_k|^2}{\left(\sum_{k=1}^n |\beta_k|^2\right)^2} \in [0,1]$ (see Methods). In the third step all the null pairs (s,t) that represent the entropy of the swarm are determined.

Null pairs can then be split up (i.e. *burned*) or left untouched. The probability for splitting up a given null pair is ϵ_j^2 hence characterizing the burn rate. In that case s stays with the swarm and t leaves it or vice versa. The probability distribution for s to leave or to stay is, in most cases, only slightly distorted towards one or the other. More specifically, the probability for s to stay is $\frac{\epsilon_j(S\setminus\{s\})}{\epsilon_j(S\setminus\{s\})+\epsilon_j(S\setminus\{t\})}$. This rule sets a weak trend towards lower excitations. The process stops as soon as all the excitations values become zero, i.e. the swarm reached a final decision.

I implemented the GenI process in JAVA (see data availablity section) and generated all the statistical data out of this reference implementation.

2.2 QM measurement statistics

Looking at the chaotic behaviour of the GenI decision making process as documented in figure 4, it would be counter-intuitive to expect any well defined statistics regarding the process outcome.

However, GenI's few quite simple rules lead the swarm state to one of the given environment options a_j with a surprisingly well defined probability distribution. This distribution exactly matches the well known predictions of quantum measurements on a physical particle given by $\frac{|\beta_j|^2}{\sum_{k=1}^n |\beta_k|^2}$. Also all the characteristics of sequential spin measurements with varying observables (e.g. different space directions) are fully met by the model (see methods section). Importantly, these remarkable characteristics are intrinsic to the GenI process, i.e. they occur without any requirement for parameter tuning.

Performing simulations on P-swarms clearly support the hypothesis, that the frequencies are indeed produced by a probablility distribution according to quantum measurements on spin½ particles (Figure 5). Equally, simulations on E-swarms demonstrate, that the observed frequencies get produced by a probability distribution $p_j = \frac{|\beta_j|^2}{\sum_{k=1}^{n}|\beta_k|^2}$ according to swarm amplitudes $S = \sum \beta_j a_j$ (table 1).

Tests have been performed for one up to millions of swarm members, starting with tons of entropy or no null pairs at all. The statistics robustly meet quantum measurement predictions producing chi square test values within 95% confidence. Swarm sizes are ultimately restricted only by the hardware capabilities that perform the simulation.

The decision process evolves completely chaotic in every aspect. This is true for the paths of amplitudes in the complex plane (Figure 6c), as well as for the evolution of absolute amplitudes over time (Figure 6a). These findings demonstrate the unpredictable nature of the Genl decision process.

The picture is similar for E-swarms and P-swarms with a slightly different pattern only for the entropy evolution (Figure 7).

2.3 Spacetime metric

General relativity is the world of continuity, unrestricted predictability of the past and the future, differentiable manifolds, four-dimensional spacetime and an reversible timeline. In sharp contrast to this, the Genl process is jumpy, discrete, chaotic, completely unpredictable, multidimensional with a clear unidirectional timeline.

However statistically the chaos can be averaged out and the discreteness of the process steps blurs with the increasing number of iterations. The resulting dynamics indeed satisfy Einstein's field equation as will be shown. To make meaningful progress at this point, one has to consider GRT as being a statistical theory and that the time is not a continuous, but a discrete variable. Several authors have already suspected this before[18].

Now taking the iteration count as a time variable, then for one single burn step we get

$\frac{db_j}{dt} \sum_{k=1}^{n} b_k^2 = \frac{1}{b_j} \left(b_j^2 - \sum_{k \neq j} b_k^2 \right)$ where $b_j = |\beta_j|$.

We will focus here on the decision process with only two choices operating on an E-swarm. In this case we get simply $\frac{dx}{dt} = \frac{1}{x} \frac{x^2 - y^2}{x^2 + y^2}$ and

$\frac{dy}{dt} = \frac{1}{y}\frac{y^2-x^2}{x^2+y^2}$ for a swarm $S = \beta_1 a_1 + \beta_2 a_2$ where $\beta_1 = xe^{i\phi_x}$, $\beta_2 = ye^{i\phi_y}$.

A map into a 4-dimensional manifold is chosen as $\vec{X} = \begin{bmatrix} x_0 \\ x_1 \\ x_2 \\ x_3 \end{bmatrix} =$

$\begin{bmatrix} t \\ \langle p_1 S \mid S \rangle \\ \langle p_2 S \mid S \rangle \\ \langle p_3 S \mid S \rangle \end{bmatrix} = \begin{bmatrix} t \\ 2cxy \\ 2sxy \\ x^2 - y^2 \end{bmatrix}$ and the process time t as the path variable

where $c = \cos(\phi_x - \phi_y)$, $s = \sin(\phi_x - \phi_y)$.

Applying the Hamiltonian minimal principle to determine the coefficients in the relativistic line element $ds^2 = Adt^2 - B_1 dx_1^2 - B_2 dx_2^2 - B_3 dx_3^2$ finally results in a metric

$$g_{\mu\nu} = \frac{x_1^2+x_2^2}{x_1^2+x_2^2+x_3^2} e^{-\frac{x_3^2}{x_1^2+x_2^2}} \begin{bmatrix} 32\left(1+\frac{x_3^2}{x_1^2+x_2^2}\right) & 0 & 0 & 0 \\ 0 & -1 & 0 & 0 \\ 0 & 0 & -1 & 0 \\ 0 & 0 & 0 & -\left(2+\frac{x_3^2}{x_1^2+x_2^2}\right) \end{bmatrix}$$

that is indeed consistent with the requirements of GRT. It determines a curved spacetime so that the GenI process locally follows geodetic lines on timelike paths. When the swarm has made its decision the metric collapses to zero. The swarm's image in spacetime always ends up in a singularity. From here it is a straightforward exercise to determine the left side of Einstein's field equation

$$R_{\mu\nu} - \frac{1}{2}g_{\mu\nu}R = -\kappa T_{\mu\nu}$$

and thus conclude a mass-energy distribution that represents the right side.

3 Discussion

A ToE that unifies QM and GRT has been a long-sought goal in modern physics. Previous approaches almost exclusively focused on QM as a potential ToE entry point. This was probably due to a wide agreement within the physics community that QM would likely represent the universe's most basic operation principle and should consequently imply GRT as well as any other physical law.

In this work, I took a radically different approach by developing a theoretical model for intelligent decision making with the principle aim to better understand intelligent behaviour and consciousness as well as to advance AI implementation. To my own surprise, the presented Genl decision making model strongly suggests a fundamental relationship between intelligence, QM and GRT. More specifically, all three domains simply represent different aspects of a previously unknown, swarm-like stochastic process operating at the very basis of our universe. Gravitation, space and time as formulated in GRT are thereby specific traits of the 'inner' view of the swarm itself, while QM describes the outer view in which the 'inner' parameters of space and time are irrelevant. Hence, GRT cannot be directly derived from QM (or vice versa), as unsuccessfully tried in the past. Both theories rather represent different perspectives onto the same, more fundamental principle formalized in Genl.

Compared to the extremely complex supergravity-, string- and M-theories aiming at explaining gravity and spacetime from the quantum perspective, the simplistic Genl model is profoundly superior. It necessitates only a small set of simple assumptions and rules and does not require any further fine-tuning of model parameters to function robustly. As its underlying mathematics are basic and straight-forward, Genl is also highly preferable according to ancient Occam's razor principle[19]. Finally, Genl could be readily examinable experimentally from various angles (see below).

Therefore, my work has far-ranging, at first sight likely irritating consequences for our understanding of what makes up the universe and how it evolves. The perceivable universe appears not to be developing according to a Schroedinger equation, as usually assumed in modern physics without any explicit justification. It rather develops according to a QM measurement process, i.e. its wave function is collapsing. Importantly, this collapse is independent of any external observer or environment, as GenI's basic P-swarm process is directly self referential. The swarm is capable to observe itself, define its targets and thereby conduct the wave function collapse completely on its own. The particular role of gravitation in this wave function collapse is in determining its inner dynamical rules rather than being its initializer.

GenI may well allow abandoning dark matter, dark energy and/or gravitons, all of which look merely like artefacts of an incomplete theory and hence still resist any experimental confirmation. Similarly, the mysterious big bang may simply resolve into a change in perspective. As has been shown, the GenI process results in a singularity when reaching its very final decision. However, even the slightest change in perspective immediately triggers swarm movement, thereby creating a new space-time metric. So the process jumps out of the singularity and starts running again. Possible causes for such a perspective change remain to be explained.

From a more general viewpoint, the GenI model can help to better understand processes that may at first sight look completely random. This is true not only for the measurement process in QM as shown here, but also for the evolution of live by genetic variation and subsequent selection, the decision making in social teams and many more. In all of these areas, GenI should allow predictions readily accessible with existing technologies. Finally, provided the GenI process indeed underlies the fuzzy concepts of consciousness and intelligent behaviour, these phenomena may be much more common in our universe than we expected until today.

More recently, the subject of consciousness has indeed become the focus of scientists who discuss its significance in the universe within multidisciplinary working groups. It is possible that consciousness, if properly defined, is a fundamental property of all matter and not, as previously thought, a phenomenon that only arises from its sufficiently complex accumulation virtually out of nothing.

Common to all these initiatives is the lack of a verifiable mathematical model consistent with existing physical theories that could solidly support such a thesis. This work describes such a model that not only has the potential to reconcile the two basic physical theories, but also takes the topic of intelligent decision-making centre stage. Genl describes a fundamental change of perspective not only for physics. It basically means that gravity is nothing else than a side effect of an evolutionary selection process. Conversely, a strong selection pressure of biological systems should show up in statistical distortions, which can not be explained by known natural laws.

"It will need a profound change of viewpoint, which makes it hard to speculate on the specific nature of the change. Moreover, it will undoubtedly look crazy!" suspected Penrose 1995 in 'shadows of the mind' when reasoning about possible approaches towards a truly generalizable ToE. In 'What Remains To Be Discovered' John Maddox highlighted 1998 "As with æther, serious people hunt for the constituents of the missing mass without acknowledging that the whole idea may be no more than a sign that present understanding of the universe is incomplete, as was Maxwell's electromagnetism without relativity. (...) My hunch is that the future will follow the past in revealing a new nest of Russian dolls to be unscrewed." And 2007 in 'THE ROAD TO REALITY' Penrose realized almost resignedly "(...) I do not believe that we have yet found the true road to reality".

I do not want to claim yet, that Genl points towards this road to reality. It remains to be demonstrated, for instance, that the beautiful performance of Genl in multiple dimensions gives as well rise to a space time compatible with GRT as has been shown for two degrees of freedom. At least, however, I took the first few humble steps on a path that nobody seriously tried before. We will see where it leads us.

4 Methods

4.1 P-swarms

The Pauli group P indicates the group $P = \{p_i\}_{i=0...15}$ that is generated by the algebraic features of the well known Pauli matrices. By embedding this group within its common irreducible representation $A_P \subsetneq Mat(2 \times 2, \mathbb{Z}[i]) \subsetneq Mat(2 \times 2, \mathbb{C})$ the group generators will often be referred to as $p_1 = \begin{bmatrix} 0 & 1 \\ 1 & 0 \end{bmatrix}, p_2 = \begin{bmatrix} 0 & -i \\ i & 0 \end{bmatrix}, p_3 = \begin{bmatrix} 1 & 0 \\ 0 & -1 \end{bmatrix}$, p_0 representing the identity, and in general $P = \{p_{j+4k} = i^k p_j | j, k = 0...3\}$ are all the respective group elements. A_P is not the full integer matrix algebra because e.g. $\begin{bmatrix} 1 & 0 \\ 0 & 0 \end{bmatrix} \notin A_P$ but $\begin{bmatrix} 2 & 0 \\ 0 & 0 \end{bmatrix} \in A_P$. Such differences may cause some mathematical headache and should get relevant for a fundamental analysis of the GenI model but are not important here. Especially for large swarms, their discrete nature will be blurred almost to a continuum.

Definition 1 *Let P be the Pauli group, E a countable set, $\rho : E \to P$ a surjective map, and $\forall p \in P : |\rho^{-1}(\{p\})| = \infty$. (E, ρ) is called the **P pool of entities**. The image $\rho(s)$ is called the type of s.*

*A **P-swarm** S is a finite subset of entities $S = \{s_j \mid j = 1 \ldots N\} \subset E$. By defining $\tilde{S} := \sum \rho(s_j) \in A_P$ S itself receives an image in the algebra. The mapping $\{S \subset E : |S| < \infty\} \to A_P$ again is surjective.*

If S and U are swarms then $\widetilde{S \cup U} = \tilde{S} + \tilde{U} - \widetilde{S \cap U}$.

*A pair $(s, t) \in E^2$ is called a **null pair**, if $\rho(s) + \rho(t) = 0$.*

*A tuple $(s_0, s_1, s_2, s_3) \in E^4$ is called a **null ring generated by** s_0, if $\exists k : \rho(s_j) = i^j p_k$*

*A swarm $N \subset E$ with $\tilde{N} = 0$ is called a **null swarm**.*

*A **maximal null swarm** $N \subset S$ is a null swarm that cannot get extended by another null pair.*

Though such a maximal null swarm is not unique the differences are not important for most practical purposes. So N_S may denote the **entropy** of the swarm. In the following I will not always distinguish between the elements and their images in A_P wherever it gets clear from the context what is meant.

Definition 2 *A **perspective** on A_P is a vector $v \in \mathbb{C}^2$ mapping $\tilde{S} \mapsto \tilde{S}v$. Sv is called the **state of S under the perspective** v.*

*A base $\{a_1, a_2\} \subset \mathbb{C}^2$ is called an **environment** identifying the possible choices the swarm may take.*

*$Sv = \beta_1 a_1 + \beta_2 a_2$ is called the **decomposition of S with respect to the environment** $\{a_1, a_2\}$.*

*$\beta_j a_j$ is called an **idea of S** about $\{a_1, a_2\}$ and $\beta_j \in \mathbb{C}$ is called its **amplitude**. So the state of S is a superposition of ideas about an environment.*

Such an environment may get determined by an operator in terms of its eigenvectors. In QM such an operator is called an observable and can be derived from the swarms image in A_P itself. So the environment need not be an external construct.

4.2 The GenI process on two options

I start with some question that allows exactly two answers - say YES or NO. These choices get represented by the environment $\{a_1, a_2\} \in \mathbb{C}^2$ where any $\gamma_1 a_1$ should mean YES and any $\gamma_2 a_2$ means NO.

Definition 3 *Let $Sv = \beta_1 a_1 + \beta_2 a_2$ be a swarm with its decomposition according to a given environment.*

*$\epsilon := \sin(\Phi) = 2\frac{|\beta_1||\beta_2|}{\beta_1^2 + \beta_2^2}$ is called the **excitation of the swarm**.*

*The angle $\Phi \in [0, \pi]$ is defined by $\cos(\frac{\Phi}{2}) = \frac{|\beta_1|}{\beta_1^2 + \beta_2^2}$ and called the **state angle**.*

The random walk will minimize the excitation
$\epsilon \to 0$ leaving $\cos(\Phi) \to \pm 1$.

I am going to design a decision process on a P-swarm S that definitely gives one of these answers randomly but with a definite probability distribution that only depends on the starting state of S.

Its decomposition into $Sv = \beta_1 a_1 + \beta_2 a_2$ holds a superposition of both answers. There is a separation of $S = S_D + N_S$ into a null swarm N_S and the rest. The null swarm is completely undecided about the given options whereas $S_D v = Sv$ completely determines the state.

The following process actually starts a competition between the two ideas with exactly defined win chances for either outcome. Regardless how small an amplitude is it will have a reasonable win opportunity. The decision process should arrive at YES-NO with probability $\frac{|\beta_i|^2}{\beta_1^2+\beta_2^2}$ respectively.

Definition 4 *Let S be an arbitrary P-swarm, $(a_1, a_2) \in \mathbb{C}^2$ an environment, $S = S_D + N_S$ a decomposition of the swarm into a maximal null swarm N_S and the entropy free rest S_D, v a given perspective and $Sv = \beta_1 a_1 + \beta_2 a_2$ its unique decomposition.*

Let $S^{(n)} v = \beta_1^{(n)} a_1 + \beta_2^{(n)} a_2$, $S^{(n)} = S_D^{(n)} + N_S^{(n)}$ be a series of P-swarms and

$$\epsilon^{(n)} = \epsilon(S^{(n)}) = 2\frac{\left|\beta_1^{(n)}\right|\left|\beta_2^{(n)}\right|}{\left|\beta_1^{(n)}\right|^2+\left|\beta_2^{(n)}\right|^2}$$

be the respective excitations and $k^{(n)} = \frac{1}{2}\left|N_S^{(n)}\right|$ the number of null pairs.

The series starts at $n = 0$ with a given $S^{(0)} = S$ containing $k^{(0)}$ null pairs.

A Genl process is a stochastic process defined by the iteration as follows:

Step 0 $n := 0$

Step 1 *Each element $s \in S_D^{(n)}$ acquires a null ring generated by s.*

Step 2 *Each null pair (r, t) randomly gets burned with probability $p = \epsilon^{(n)2}$. For each burned null pair the element r gets randomly chosen to stay with the swarm with probability*
$p(r stays) = \frac{\epsilon(S+t)}{\epsilon(S+r)+\epsilon(S+t)}$, *otherwise t stays and r leaves. By following this rule the system tends to reduce ϵ.*

Step 3 *The remaining null pairs stay with the swarm. The resulting swarm is $S^{(n+1)}$.*

Step 4 *If $\epsilon^{(n+1)} = 0$ stop. Otherwise iterate $n \to n+1$ and continue with Step 1.*

Without knowing anything about the dynamic rules each burn step would certainly look uniformly random. The probability distortion is hard to detect unless the system gets close to one of the options defined by its environment. This is due to the fact that $\epsilon(S+r)$ only slightly differs from $\epsilon(S-r)$ especially for large swarms of probably millions of members.

4.3 E-swarms

In order to enable more than two choices two or more P-swarm images can get combined to form a higher order tensor algebra. Such a construct relates to a swarm composed of strings of P-swarm individuals. A higher order P-Swarm then is simply a set of such strings. Basically this is exactly where to look for.

Given two swarms S and U the union $S \cup U$ is simply a bigger swarm whereas the set product $S \times U$ is certainly something else. With respect to the algebraic image we can define a map $\rho : S \times U \to A_P \otimes A_P$ by $\rho(s, u) = \rho(s) \otimes \rho(u)$ leading directly to $\rho(S \times U) = \sum_{\substack{s \in S \\ u \in U}} \rho(s) \otimes \rho(u)$.

Given an array of N swarms $S_1, \ldots, S_N$ we define $\rho : S_1 \times \ldots \times S_N \to \otimes_1^N A_P$ by $\rho(s_1, \ldots, s_N) = \rho(s_1) \otimes \ldots \otimes \rho(s_N)$ the ordinary tensor product and for any $U \subset S_1 \times \ldots \times S_N = S$ we get $\rho(U) = \sum_{(s_i)_{i=1\ldots N} \in U} \rho(s_1) \otimes \ldots \otimes \rho(s_N)$.

It is now clear how high order swarms may map to any high order tensor algebra. Again we have a canonical base $(p_{i_1} \otimes \ldots \otimes p_{i_N} : i_k \in \{0; 1; 2; 3\})$ of $\mathbb{C}$-dimension 4^N, respectively 2^N after applying a perspective $v \otimes \ldots \otimes v$. Such a high order algebra gives rise to more simplifications. Well known from QM is a symmetry transformation by considering the elements in $(s_1, \ldots, s_N)$ indistinguishable. Summarizing over all Permutations reduces the dimension finally down to $N+1$.

This representation only considers the number of YES' from $0\ldots N$ as allowed environment options, while disregarding who specifically said YES and who answered NO. In QM each of these numbers typically relates to an eigenvector of a suitable observable.

So a complementary swarm concept gets introduced here. It simplifies many of the problems to be expected otherwise and in case of two dimensions it is equivalent to the model above with respect to the statistics produced by the GenI process.

Definition 5 Let $B = \{a_j | j = 1\ldots n\} \subset \mathbb{C}^n$ be the canonical base representing the possible choices, E a countable set, $\rho : E \to \tilde{B} = \{i^k a : k = 0\ldots 3, a \in B\}$ a surjective map, and $\forall a \in \tilde{B} : |\rho^{-1}(\{a\})| = \infty$. (E, ρ) is called the **E pool of entities**. The image $\rho(s)$, $s \in E$, is called the type of s.

A **E-swarm** S is a finite subset of entities $S = \{s_j \mid j = 1\ldots N\} \subset E$. By defining $\tilde{S} := \sum \rho(s_j) \in \mathbb{C}^n$ S itself receives an image in the vector space.

A pair $(s,t) \in E^2$ is called a **null pair**, if $\rho(s) + \rho(t) = 0$.
A tuple $(s_0, s_1, s_2, s_3) \in E^4$ is called a **null ring generated by** s_0, if $\exists k : \rho(s_j) = i^j e_k$.

A swarm $N \subset E$ with $\tilde{N} = 0$ is called a **null swarm**.

A **maximal null swarm** $N \subset S$ is a null swarm that cannot get extended by another null pair. Such a sub-swarm is called the **entropy of the swarm**.

$S = \sum_{j=1}^{n} \beta_j a_j$ is called the **decomposition of S with respect to B**. $\beta_j a_j$ is called an **idea of S about B** and $\beta_j \in \mathbb{C}$ is called its **amplitude**. If $\beta_j = 0$ then it is said that S has no idea about option a_j.

So any such swarm can be viewed as a superposition of ideas about a subset of options given by the environment. Here each individual is completely decided about one of the options.

This level of simplification, however, does not allow a change of the environment. To enable that, one option is to reorganize basic swarm members into indivisible atoms that represent the vectors in $\tilde{B}$. So changing the set of options reorganizes the swarm into a new set of atoms as a first step, where each atom again is completely decided about one of the new options. This can happen under a few obvious conditions, but the details are not as simple as they appear at first glance and the additional complexity is not required for the purpose of this work. Above all, such an approach will not change the statistics of the GenI process, as defined below.

4.4 The GenI process on E-swarms

To extend the GenI process to multiple choices I think of a swarm as separated into $S = \beta_j a_j + \sum_{k \neq j} \beta_k a_k = \beta_j a_j + \gamma_j u_j$ for each index j where u_j is orthogonal to a_j. If u_j is a normalized vector then $|\gamma_j| = \sqrt{\sum_{k \neq j} |\beta_k|^2}$.

Definition 6 $\epsilon_j := \sin(\Phi_j) = 2 \frac{|\beta_j| \sqrt{\sum_{k \neq j} |\beta_k|^2}}{|S|^2}$ is called the **excitation of idea** j.

The **state angles** $\Phi_j \in [0, \pi]$ are defined by
$\cos(\frac{\Phi_j}{2}) = \frac{|\beta_j|}{|S|}$, equivalent to $\sin(\frac{\Phi_j}{2}) = \frac{\sqrt{\sum_{k \neq j} |\beta_k|^2}}{|S|}$.

Now everything else is indeed straightforward.

Definition 7 Let S be an arbitrary E-swarm, $B = \{a_j\}_{j=1...n} \subset \mathbb{C}^n$ an environment, $\rho : S \to \{i^k a : a \in B\}$ the map into the vector space, $S = S_D + N_S$ a decomposition of the swarm into a maximal null swarm N_S and the rest, $S = \sum_{k=1}^{n} \beta_k a_k, \beta_k \in \mathbb{C}$ its unique decomposition, $b_j = |\beta_j|$ and $y_j = \sqrt{\sum_{k \neq j} |\beta_k|^2}$.

Let $S^{(n)} = \sum_{k=1}^{n} \beta_k^{(n)} a_k$, $b_k^{(n)} = \left|\beta_k^{(n)}\right|$, $y_j^{(n)} = \sqrt{\sum_{k \neq j} \left|\beta_k^{(n)}\right|^2}$, $S^{(n)} = S_D^{(n)} + N_S^{(n)}$ be a series of E-swarms and $\epsilon_j^{(n)2} = \epsilon_j(S^{(n)})^2 = 4 \frac{b_j^{(n)2} y_j^{(n)2}}{|S^{(n)}|^4}$ be the respective **excitations**.

The series starts at $n = 0$ with a given $S^{(0)} = S$.

A **GenI process** is a stochastic process defined by the iteration as follows:

Step 0 $n := 0$

Step 1 Each element $s \in S_D^{(n)}$ acquires a null ring generated by s.

Step 2 Each null pair (r,t), $\tilde{r} = i^k a_j, \tilde{t} = -i^k a_j$ randomly gets burned with probability $p = \epsilon_j^{(n)2}$. For each burned null pair the element r gets randomly chosen to stay with the swarm with probability $p(rstays) = \frac{\epsilon_j(S+t)}{\epsilon_j(S+r)+\epsilon_j(S+t)}$, otherwise t stays and r leaves. By following this rule the system tends to reduce ϵ_j.

Step 3 The remaining null pairs stay with the swarm. The resulting swarm is $S^{(n+1)}$.

Step 4 If $\forall j = 1 \ldots n : \epsilon_j^{(n+1)} = 0$ then stop. Otherwise iterate $n \to n+1$ and continue with Step 1.

A sample statistic is given in figure 1. The target values represent the expected probabilities given by $p_j = \frac{b_j^2}{\sum b_k^2}$. Sigmas are calculated from target probabilities by $\sigma_j^2 = \frac{1}{n} p_j (1 - p_j)$.

4.5 Transformations

To be relevant for QM, the transformation behaviour has to be considered when switching the environment. Basically the GenI model meets the typical QM transformations by design. So changing the environment provides a transformation adapting the amplitudes in exactly the way, that a change of QM observable would do.

To demonstrate that in case of YES-NO type decision, let an orthonormal base $a_1, a_2 \in \mathbb{C}^2$ be the chosen environment. The Pauli matrix p_3 is the typical observable for QM spin measurements along a z-axis in 3-dimensional space. Then (a_1, a_2) are its eigenvectors with eigenvalues ± 1.

Let $Sv = xe^{i\phi_x}a_1 + ye^{i\phi_y}a_2$, $x,y \in \mathbb{R}_0^+$, $s = \sin(\phi_x - \phi_y)$, $c = \cos(\phi_x - \phi_y)$, and $\vec{r}(Sv) = \begin{bmatrix} x_1 \\ x_2 \\ x_3 \end{bmatrix} = \begin{bmatrix} \langle p_1 Sv \mid Sv \rangle \\ \langle p_2 Sv \mid Sv \rangle \\ \langle p_3 Sv \mid Sv \rangle \end{bmatrix} = \begin{bmatrix} 2cxy \\ 2sxy \\ x^2 - y^2 \end{bmatrix}$

Then the GenI process excitation $\epsilon^2 = x_1^2 + x_2^2 \to 0$ and $x_3 \to \pm r$. Now spin-up/-down along the z-axis is indeed represented by a final state given by $\vec{r} = \begin{bmatrix} 0 \\ 0 \\ \pm r \end{bmatrix}$.

Switching the observable to p_1 and hence the environment to the according eigenvectors (a_1', a_2') with eigenvalues ± 1 gives $Sv = x' e^{i\phi'_x} a_1' + y' e^{i\phi'_y} a_2'$ and $\vec{r}' = \begin{bmatrix} x'^2 - y'^2 \\ 2c'x'y' \\ 2s'x'y' \end{bmatrix}$ Again the behaviour meets the conditions of QM spin measurement along the x axis.

In general I may choose an arbitrary space direction for spin measurement. It is well known, that the formal Pauli vector (p_1, p_2, p_3) transforms exactly as the three dimensional axes do.

Finally it looks like I could define the GenI process also in terms of the special mapping in $\mathbb{R}^3$. But that is not true. There is no way, for example, to represent the swarm entropy. Splitting up a null pair leads to two identical vectors $\vec{r}(Sv) = \vec{r}(-Sv)$.

4.6 Restarting the process

Having made a decision about an environment B the GenI process finally stops. At this point the swarm's state matches exactly one of the given options with no idea left about all the others. If we present B again nothing will happen and the swarms sticks to its previous decision according to definition 7.

Now change the environment to B'. The swarm state that previously got definite about an option of B now may be completely open with respect to B'. The new amplitudes are well defined by the transformation $B \to B'$. So the process newly starts over to reach the according answer to the new question. If B is presented again, then the swarm can be completely undecided now, without any memory of its former state, and come to a different decision than the first time. Due to the design of the Genl model, this behaviour copies exactly the conditions that are found in quantum mechanics when changing the observable after a completed measurement. Such a case occurs, for example, when measuring the spin of an electron once in the z-direction, then in the x-direction, and finally again in the z-direction.

A P-swarm S may act according to its internal environment under a given perspective v. Let A_S be an internal observable derived from the operator image of S, whose eigenvectors determine an environment B_S. For example $A_S = \frac{1}{2}(S+S^\dagger)$ would do, or any other suitable linear operator with a well defined functional relationship to S. As soon as it comes to a final decision about one of the two (now moving) options the process stops according to definition 4. At that time Sv has arrived at an eigenvector of A_S. Now any change of the perspective $v \to v'$ obviously restarts the Genl process immediately, because A_S stays unchanged while Sv' in general no longer represents one of its eigenvectors.

4.7 Process statistics

The Genl process according to definitions 7 and 4 delivers extremely accurate results compared to corresponding quantum measurement statistics. There are only few cases where it does not come to a definite end after a reasonable iterations count (Table 1 / Figure 5).

These results strongly support the following (see data availability section):

Hypothesis 1 *Given a swarm S of order n with $\tilde{S} = \sum_1^n \beta_j a_j$, $b_j = |\beta_j|$, a Genl process G iterating on S resulting in a series $S^{(m)} = \sum_1^n \beta_j^{(m)} a_j$ of stochastic variables.*

Then $P\left(S^{(m)} \xrightarrow[m \to \infty]{} \gamma a_j\right) = P\left(\sum_{k \neq j} b_k^{(m)2} \xrightarrow[m \to \infty]{} 0\right) = \frac{b_j^2}{|S|^2}.$

Let us now take a statistical view on the process in case of large amplitudes. Choose $S = \sum_{j=1}^n \beta_j a_j$, $\beta_j = c_j + id_j$ and n_j null pairs according to each idea.

Step 1 in definition 7 implies that for each j exactly $2(|c_j|+|d_j|)$ null pairs get acquired, half of them potentially altering the real part and the imaginary part respectively.

Step 2 results in burning at average $\epsilon_j^2(2|c_j|+2|d_j|+n_j)$ of the old and the newly acquired null pairs and leaving $n'_j = (1-\epsilon_j^2)(2|c_j|+2|d_j|+n_j)$ of them to the next iteration.

Now look at a single burn event.

The probability to increase c_j by ± 1 is

$$P(\Delta c_j = \pm 1) = \frac{\epsilon_j(c_j \mp 1)}{\epsilon_j(c_j-1)+\epsilon_j(c_j+1)} \quad (1)$$

where the target value is $\epsilon_j = 2\frac{b_j\sqrt{\sum_{k\neq j}b_k^2}}{\sum_{k=1}^{n}b_k^2}$.

Hence the observed distribution in most cases is nearly uniform $P(\Delta c_j = \pm 1) \approx \frac{1}{2}\left(1 \mp \frac{c_j}{b_j^2}\right) \approx \frac{1}{2}$.

The process now results in a mean value
$mean(\Delta c_j) = \frac{\epsilon_j(c_j-1)-\epsilon_j(c_j+1)}{\epsilon_j(c_j-1)+\epsilon_j(c_j+1)} \approx -\frac{1}{\epsilon_j}\frac{\partial \epsilon_j}{\partial c_j} = -\frac{\partial \ln(\epsilon_j)}{\partial c_j}$.

With $\ln(\epsilon_j) = \ln(2) + \ln(b_j) + \frac{1}{2}\ln\left(\sum_{k\neq j}b_k^2\right) - \ln\left(\sum_k b_k^2\right)$ we get
$\frac{\partial \ln(\epsilon_j)}{\partial c_j} = \frac{\partial \ln(\epsilon_j)}{\partial b_j}\frac{\partial b_j}{\partial c_j} = \left(\frac{1}{b_j} - 2\frac{b_j}{\sum_k b_k^2}\right)\frac{c_j}{b_j}$ and hence
$mean(\Delta c_j) = -\frac{c_j}{b_j^2}\left(1 - 2\frac{b_j^2}{\sum_k b_k^2}\right)$
$= -\frac{c_j}{b_j^2}\left(\frac{\sum_{k\neq j}b_k^2}{\sum_k b_k^2} - \frac{b_j^2}{\sum_k b_k^2}\right)$
$= \frac{c_j}{b_j^2}\left(\cos(\frac{\Phi_j}{2})^2 - \sin(\frac{\Phi_j}{2})^2\right)$

where $\cos(\frac{\Phi_j}{2}) := \frac{|\langle S|a_j\rangle|}{|S|} = \frac{b_j}{\sqrt{\sum_k b_k^2}}$.

The same relation holds for the imaginary part d_j.

On average is $\Delta b_j = |\beta_j + \Delta\beta_j| - |\beta|_j$
$= \sqrt{(c_j+\Delta c_j)^2 + (d_j+\Delta d_j)^2} - \sqrt{c_j^2 + d_j^2}$

$\approx \frac{c_j}{b_j}\Delta c_j + \frac{d_j}{b_j}\Delta d_j$ for each amplitude. So

$$mean(\Delta b_j) = \frac{1}{b_j}\left(\cos(\frac{\Phi_j}{2})^2 - \sin(\frac{\Phi_j}{2})^2\right) = \frac{1}{b_j}\cos(\Phi_j) \in [-\frac{1}{b_j}; \frac{1}{b_j}] \quad (2)$$

for each single burn event.

Now take the number of burn steps as time axis leading to a system of nonlinear differential equations that should approximately describe the process dynamics at a statistical level:

$$\frac{db_j}{dt}\sum_{k=1}^{n} b_k^2 = \frac{1}{b_j}\left(b_j^2 - \sum_{k \neq j} b_k^2\right) \quad (3)$$

Such a single burn step given in equations (2) and (3) basically exposes some sort of a force overlaying the random movements and leading to a probability distortion toward selected directions.

As you see from equation 3 any b_j tends to increase the according amplitude only if its square is larger than the sum of all the others. Otherwise it tends to decrease further. This effectively limits the swam amplitudes and avoids any catastrophic behaviour for almost all of the process time. Additionally the change rate (and hence the probability distortion) is tiny for high amplitudes in the order of $\frac{1}{b_j}$.

4.8 A spacetime geometry

I will focus here on the decision process on an E-swarm S with only two choices and make some definitions to improve readability.

The swarm is decomposed as $S = \beta_1 a_1 + \beta_2 a_2$ where $\beta_1 = xe^{i\phi_x}, \beta_2 = ye^{i\phi_y}$, $b = |S| = \sqrt{x^2 + y^2}$.

Equation 3 gives $\frac{dx}{dt} = \frac{1}{x}\frac{x^2-y^2}{x^2+y^2}$ and $\frac{dy}{dt} = \frac{1}{y}\frac{y^2-x^2}{x^2+y^2}$ and hence $\frac{d}{dt}|S|^2 = 2x\frac{dx}{dt} + 2y\frac{dy}{dt} = 0$. So the norm b of S is a process constant.

Now I have to define an appropriate metric in spacetime that allows the process dynamics to follow geodetic lines. This means I have to find real functions A, B_1, B_2, B_3 so that the path length $\int ds$ gets stationary with the line element

$$ds^2 = A dx_0^2 - B_1 dx_1^2 - B_2 dx_2^2 - B_3 dx_3^2 \qquad (4)$$

Now choose the map $\vec{X} = \begin{bmatrix} x_0 \\ x_1 \\ x_2 \\ x_3 \end{bmatrix} = \begin{bmatrix} t \\ \langle p_1 S \mid S \rangle \\ \langle p_2 S \mid S \rangle \\ \langle p_3 S \mid S \rangle \end{bmatrix} = \begin{bmatrix} t \\ 2cxy \\ 2sxy \\ x^2 - y^2 \end{bmatrix}$ and the process time t as the path variable where $c = \cos(\phi_x - \phi_y)$, $s = \sin(\phi_x - \phi_y)$.

Let $r = \sqrt{x_1^2 + x_2^2} = 2xy$.

I get $\dot{\vec{X}} = \begin{bmatrix} \dot{x}_0 \\ \dot{x}_1 \\ \dot{x}_2 \\ \dot{x}_3 \end{bmatrix} = \begin{bmatrix} 1 \\ -4c\frac{x_3^2}{rb^2} \\ -4s\frac{x_3^2}{rb^2} \\ 4\frac{x_3}{b^2} \end{bmatrix}$

and $\ddot{\vec{X}} = \begin{bmatrix} \ddot{x}_0 \\ \ddot{x}_1 \\ \ddot{x}_2 \\ \ddot{x}_3 \end{bmatrix} = \begin{bmatrix} 0 \\ -16c\frac{x_3^2}{rb^4}\left(2+\frac{x_3^2}{r^2}\right) \\ -16s\frac{x_3^2}{rb^4}\left(2+\frac{x_3^2}{r^2}\right) \\ 16\frac{x_3}{b^4} \end{bmatrix}$

Now I am going to determine the functions A, B_j. The line element 4 leads to the Lagrange term $L = \frac{1}{2}\left[A\dot{x}_0^2 - \sum_j B_j \dot{x}_j^2\right]$ according to the Hamiltonian extremal principle. The Euler-Lagrange equations $\frac{d}{dt}\frac{\partial L}{\partial \dot{x}_j} - \frac{\partial L}{\partial x_j} = 0$ finally result in

$$2\frac{\partial A}{\partial \dot{x}_0}\dot{x}_0\ddot{x}_0 + 2A\ddot{x}_0 - 2\sum_{k=1}^{3}\frac{\partial B_k}{\partial \dot{x}_0}\dot{x}_k\ddot{x}_k - \frac{\partial A}{\partial x_0}\dot{x}_0^2 + \sum_k \frac{\partial B_k}{\partial x_0}\dot{x}_k^2 = 0 \quad (5)$$

and for $j = 1 \ldots 3$

$$2\frac{\partial A}{\partial \dot{x}_j}\dot{x}_0\ddot{x}_0 - 2\sum_k \frac{\partial B_k}{\partial \dot{x}_j}\dot{x}_k\ddot{x}_k - 2B_j\ddot{x}_j - \frac{\partial A}{\partial x_j}\dot{x}_0^2 + \sum_k \frac{\partial B_k}{\partial x_j}\dot{x}_k^2 = 0 \quad (6)$$

The basic symmetry of the process space suggests $B_1 = B_2 = B(\frac{r}{x_3}), B_3 = C(\frac{r}{x_3}), A = A(\frac{r}{x_3})$. So all functions do not depend on x_0 and the $\dot{x}_j$. Then 5 is identical zero and the above terms now get inserted into the remaining geodetic equations 6 :

$$0 = -2B_j \ddot{x}_j - \frac{\partial A}{\partial x_j}\dot{x}_0^2 + \sum_{k=1}^{3}\frac{\partial B_k}{\partial x_j}\dot{x}_k^2 \tag{7}$$

With $z = \frac{r}{x_3}$ I get $\frac{\partial}{\partial x_1} = \frac{c}{x_3}\frac{d}{dz}$, $\frac{\partial}{\partial x_2} = \frac{s}{x_3}\frac{d}{dz}$, $\frac{\partial}{\partial x_3} = -\frac{r}{x_3^2}\frac{d}{dz}$.
Now let's do the calculations:

$$0 = -2B_1\ddot{x}_1 - \frac{\partial A}{\partial x_j}\dot{x}_0^2 + \frac{\partial B}{\partial x_1}(\dot{x}_1^2 + \dot{x}_2^2) + \frac{\partial C}{\partial x_1}\dot{x}_3^2$$

$$= 32Bc\frac{x_3^2}{rb^4}\left(2 + \frac{x_3^2}{r^2}\right) - A'\frac{c}{x_3} + 16cB'\frac{x_3^3}{r^2b^4} + 16cC'\frac{x_3}{b^4}$$

$$= 16\frac{x_3 c}{b^4}\left(2B\frac{x_3}{r}\left(2 + \frac{x_3^2}{r^2}\right) - \frac{1}{16}A'(1 + \frac{r^2}{x_3^2}) + B'\frac{x_3^2}{r^2} + C'\right)$$

and

$$0 = -2B_3\ddot{x}_3 - \frac{\partial A}{\partial x_3}\dot{x}_0^2 + \frac{\partial B}{\partial x_3}(\dot{x}_1^2 + \dot{x}_2^2) + \frac{\partial C}{\partial x_3}\dot{x}_3^2$$

$$= -32C\frac{x_3}{b^4} + \frac{r}{x_3^2}A' - 16\frac{r}{x_3^2}B'\frac{x_3^4}{r^2b^4} - 16\frac{r}{x_3^2}C'\frac{x_3^2}{b^4}$$

$$= -16\frac{r}{b^4}\left(2C\frac{x_3}{r} - \frac{1}{16}(1 + \frac{r^2}{x_3^2})A' + B'\frac{x_3^2}{r^2} + C'\right)$$

After substituting with z we find

$$2B\frac{1}{z}\left(2 + \frac{1}{z^2}\right) - \frac{1}{16}(1 + z^2)A' + B'\frac{1}{z^2} + C' = 0 \tag{8}$$

$$2C\frac{1}{z} - \frac{1}{16}(1 + z^2)A' + B'\frac{1}{z^2} + C' = 0 \tag{9}$$

so that $C(z) = B(z)\left(2 + \frac{1}{z^2}\right)$, $C'(z) = B'(z)\left(2 + \frac{1}{z^2}\right) - 2B\frac{1}{z^3}$ and with equation 8

$0 = \frac{4}{z}B + 2B'\left(1 + \frac{1}{z^2}\right) - \frac{1}{16}(1 + z^2)A'$

We have to make sure that the process follows timelike paths. That means $\dot{s}^2 = A - B(\dot{x}_1^2 + \dot{x}_2^2) - C\dot{x}_3^2 \geq 0$.

We have $\dot{x}_1^2 + \dot{x}_2^2 = 16\frac{x_3^4}{r^2 b^4} \leq 16\frac{x_3^2}{r^2}$ and $\dot{x}_3^2 = 16\frac{x_3^2}{b^4} \leq 16$.

The relation

$A \geq 16\frac{x_3^2}{r^2}B + 16C = 16\frac{x_3^2}{r^2}B + 16B\left(2 + \frac{x_3^2}{r^2}\right) = 32B\left(\frac{x_3^2}{r^2} + 1\right)$

suggests to choose $A = 32B\left(\frac{x_3^2}{r^2} + 1\right) = 32B\left(1 + \frac{1}{z^2}\right)$.

Now $0 = \frac{4}{z}B + 2B'\left(1 + \frac{1}{z^2}\right) - 2(1 + z^2)\left(B'\left(1 + \frac{1}{z^2}\right) - \frac{2}{z^3}B\right) = \frac{4}{z}\left(2 + \frac{1}{z^2}\right)B - 2(1 + z^2)B'$ leads to

$B = k\frac{z^2}{1+z^2}e^{-\frac{1}{z^2}}$,

$C(z) = k\frac{2z^2+1}{1+z^2}e^{-\frac{1}{z^2}}$, and

$A = 32ke^{-\frac{1}{z^2}}$

The metric now looks like

$M = \frac{z^2}{1+z^2}e^{-\frac{1}{z^2}}\begin{bmatrix} 32\left(1 + \frac{1}{z^2}\right) & 0 & 0 & 0 \\ 0 & -1 & 0 & 0 \\ 0 & 0 & -1 & 0 \\ 0 & 0 & 0 & -\left(2 + \frac{1}{z^2}\right) \end{bmatrix}$.

Theorem 1 *Let* $S = \beta_1 a_1 + \beta_2 a_2$ *a E-swarm,* $\beta_1 = xe^{i\phi_x}$, $\beta_2 = ye^{i\phi_y}$

and $\vec{X} = \begin{bmatrix} x_0 \\ x_1 \\ x_2 \\ x_3 \end{bmatrix} = \begin{bmatrix} t \\ 2cxy \\ 2sxy \\ x^2 - y^2 \end{bmatrix}$ *a map into a 4-dimensional manifold*

where t is the Genl process time and $c = \cos(\phi_x - \phi_y)$, $s = \sin(\phi_x - \phi_y)$.

The metric tensor

$g_{\mu\nu} =$

$\frac{x_1^2+x_2^2}{x_1^2+x_2^2+x_3^2}e^{-\frac{x_3^2}{x_1^2+x_2^2}}\begin{bmatrix} 32\left(1 + \frac{x_3^2}{x_1^2+x_2^2}\right) & 0 & 0 & 0 \\ 0 & -1 & 0 & 0 \\ 0 & 0 & -1 & 0 \\ 0 & 0 & 0 & -\left(2 + \frac{x_3^2}{x_1^2+x_2^2}\right) \end{bmatrix}$

determines a curved spacetime so that the GenI process locally follows geodetic lines on timelike paths. It has a singularity at $r = 0$ that marks the final end of the process where r collapses from a value ≥ 1 to 0.

From here it is obviously a straightforward exercise to determine the left side of Einstein's field equation

$$R_{\mu\nu} - \frac{1}{2} g_{\mu\nu} R = -\kappa T_{\mu\nu}$$

and thus conclude a mass-energy distribution that represents the right side.

Referencing definition 3 on page 16 we could write the metric in terms of the excitation $\epsilon = \frac{2xy}{x^2+y^2} = \frac{r}{b^2} = \sin(\Phi)$

$$g_{\mu\nu} = e^{-\cot(\Phi)} \begin{bmatrix} 32 & 0 & 0 & 0 \\ 0 & -\sin(\Phi)^2 & 0 & 0 \\ 0 & 0 & -\sin(\Phi)^2 & 0 \\ 0 & 0 & 0 & -(1+\sin(\Phi)^2) \end{bmatrix} \quad (10)$$

to demonstrate how the process target function governs the metric. It has singularities at $\Phi \in \{0, \pi\}$ where the process stops and the metric vanishes to zero.

The case discussed before marks the most simple situation equivalent to a measurement on one single spin½ particle. But how about many choices as represented by a swarm and environment with many dimensions?

Let $S = \sum_{j=1}^{N} \beta_j a_j$, $\beta_j = x_j e^{i\phi_j}$.

Then one suggestion is to extend the map above by $\vec{X}_j$

$$= \begin{bmatrix} 2\cos(\phi_j) x_j \sqrt{\sum_{k \neq j} x_k^2} \\ 2\sin(\phi_j) x_j \sqrt{\sum_{k \neq j} x_k^2} \\ x_j^2 - \sum_{k \neq j} x_k^2 \end{bmatrix}^t = \begin{bmatrix} 2c_j x_j y_j \\ 2s_j x_j y_j \\ x_j^2 - y_j^2 \end{bmatrix}^t.$$

This map correspond to a separation $S = \beta_j a_j + \sum_{k \neq j} \beta_k a_k = \beta_j a_j + y_j u_j$ with real y_j and a normalized vector $u_j \in \mathbb{C}(a_k)_{k \neq j}$. Applying the GenI process on S results for each index to the well known dynamics.

However performing similar calculations as before leads to a metric $g_{\mu\nu}$

$$= e^{-\cot(\Phi_j)} \begin{bmatrix} 32 & 0 & 0 & 0 \\ 0 & -\sin(\Phi_j)^2 & 0 & 0 \\ 0 & 0 & -\sin(\Phi_j)^2 & 0 \\ 0 & 0 & 0 & -(1+\sin(\Phi_j)^2) \end{bmatrix}$$

for each such index j according to equation 10. The according maps however have a different context and should not cover the same area of the manifold M. They have to be located at different points $P_j \in M$. This means that the comprehensive spacetime combines individual bubbles of local spacetime structures. Right now the extension of the model looks not like a straightforward exercise and I will certainly investigate further on that.

4.9 Data availability

The JAVA reference implementation of the GenI process is available on https://github.com/genreith/BZuS.git with sources, test data and test scripts to reproduce any of the results demonstrated here.

Software/hardware stack for development and runtime:

Hardware: AMD FX-8350 Eight-Core Processor 4.00 GHz 32 GB RAM

Operating System: Windows 7 Ultimate (64-bit) with Service Pack 1

Java runtime: Java 8 Update 121 (64-bit)

Eclipse IDE for Java Developers, Version: Neon.3 Release (4.6.3), Build id: 20170314-1500

Java SE Development Kit 8 Update 121 (64-bit)

5 Tables

Table 1: **E-swarm statistics.** Genl model output runing 1000 measurements within a 10 options environment. Statistical numbers (results) are compared to the target values (targets rounded to next integer) to be expected from quantum measurements. The chi test value 7.86 is much lower that the critical value 16.91 at 95% confidence level. The target values represent the expected probabilities given by $p_j = \frac{b_j^2}{\sum b_k^2}$. Sigmas are calculated from target probabilities by $\sigma_j^2 = \frac{1}{n} p_j (1 - p_j)$. The simulation stops after 500 iterations or if swarm size exceeds ten million or as soon as the maximal square absolute value of an amplitude gets bigger than 100 times the second largest amplitude. In the latter case that index is chosen as result. Otherwise the iterated series is considered divergent. This pragmatic approach causes only insignificant probability artefacts.

	idea 1	idea 2	idea 3	idea 4	idea 5	idea 6	idea 7	idea 8	idea 9	idea10
targets=	132	81	97	78	11	206	3	336	36	3
results=	135	74	99	76	15	189	1	357	36	1
sigma=	10.7	8.6	9.4	8.5	3.3	12.8	1.8	14.9	5.9	1.8

measurements scheduled: 1000   divergent: 17   convergent: 983

statistics:

chi test value: 7.86   chi critical at 95%: 16.91898
medium size=300418   sigma=281543   max size=1008512   min size= 9695
medium abs= 500   sigma= 471   max abs= 4084   min abs= 23

6 Figures

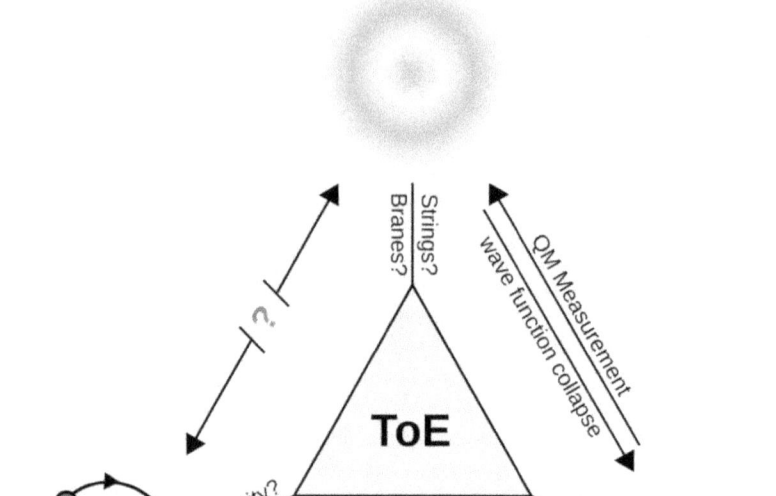

Figure 1: **Entry points to a ToE** Three at first sight very different areas: quantum mechanics, general relativity and intelligent behaviour. Each of these should be considered as an entry point for a ToE. The relationship between GRT and QM remains unclear. In QM, the observer is always part of the observation insofar as the quantum measurement itself causes the collapse of the wave function. Within GRT the observer always stands outside the system to be measured and any observation can in principle be carried out without affecting it.

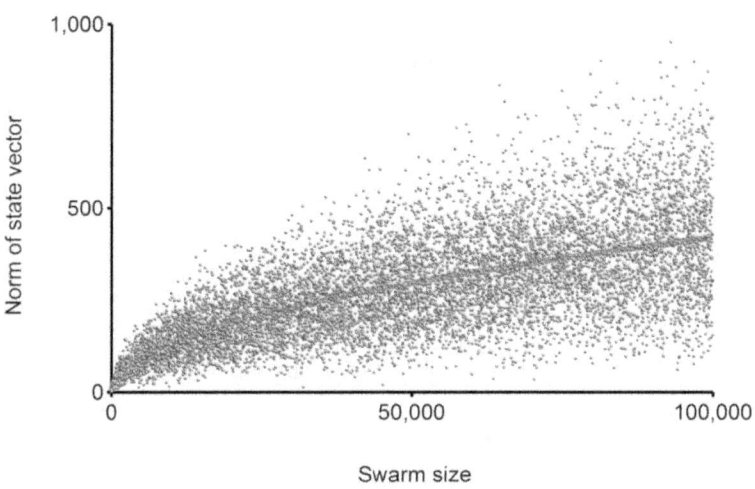

Figure 2: **Impact of swarm size on amplitudes**. Scatter chart showing swarm sizes versus state vector norms. Swarms were generated at sizes from 1 to 100,000 and randomly selected types of swarm individuals. The entropy decouples the swarm size from the norm of its state. The amplitude trend line shown is of order $\sqrt{size}$

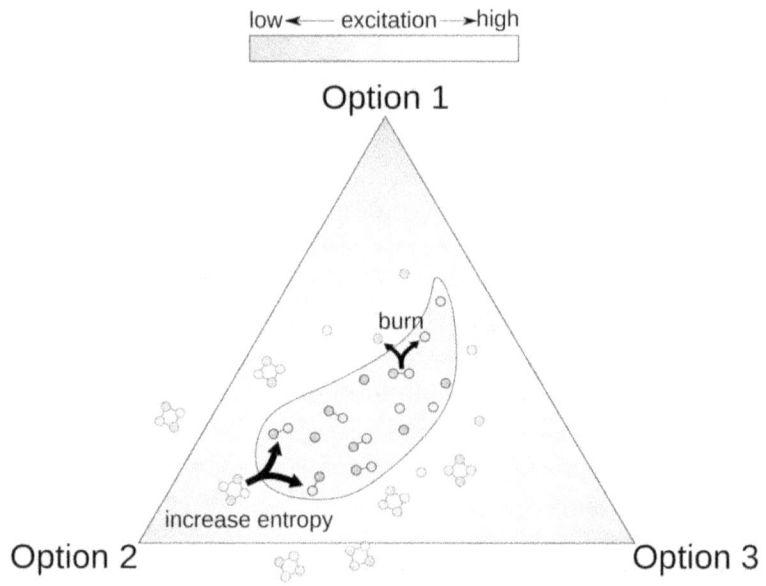

Figure 3: **The GenI process**. The GenI swarm constantly sucks null rings from its environment and burns them randomly. The GenI process sets a gradient towards reducing the excitation. The swarm's state finally arrives at one of the given options.

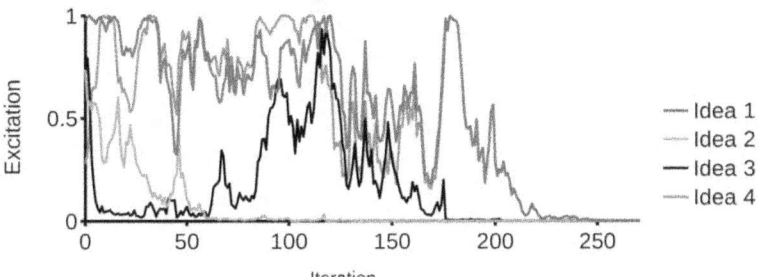

Figure 4: **Disrespecting the inner rules enables the GenI swarm to fully explore its environment.** Shown are excitation values for each idea evolving along the indicated number of iterations. The GenI process provides a gradient pointing towards lower excitation values for each idea. Due to random disrespect of the swarm's inner rules, no persistent monotony in swarm behavior does occur at any time.

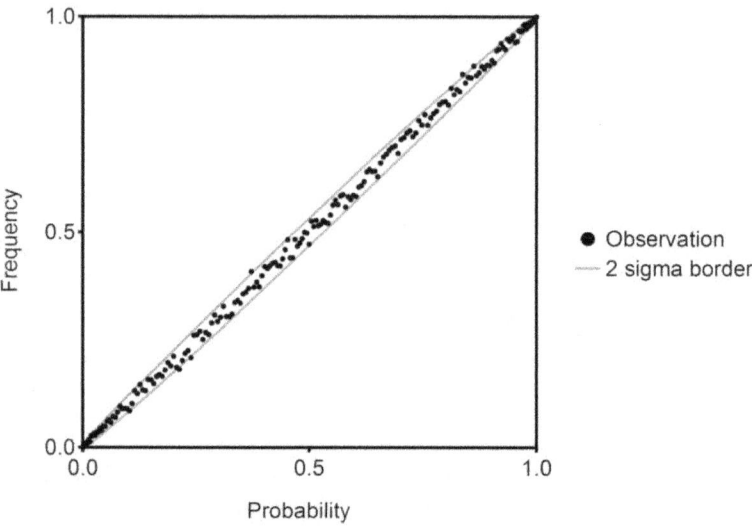

Figure 5: **P-swarms observations along target values from 0 to 1.** The sample comprises 1000 measurements each on 201 test points. More than 97% of observed frequencies are in the 2 sigma interval around the target values expected from quantum measurements. The chi square test value 92.6 is much lower than the critical level 168 at 95% confidence and 200 degrees of freedom.

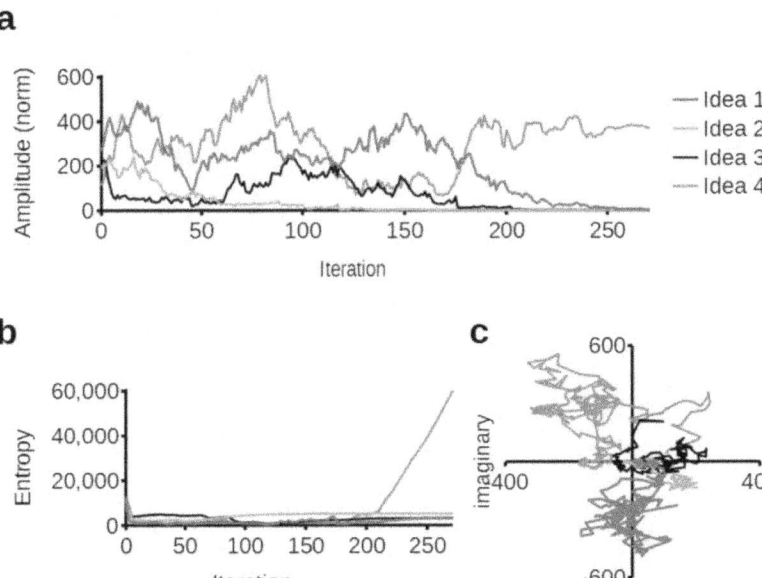

Figure 6: **Competition of ideas within an E-swarm.** Graph a shows the evolution of absolute amplitudes during a GenI process operating in a four options environment. Due to its intrinsically chaotic behavior, it is impossible to predict the GenI process evolution at any point. Interestingly, here the option with the lowest initial chance finally wins. Chart b shows the according evolution of entropy rising dramatically at the end for the winning idea. Figure c displays the native paths of each idea in the complex plane.

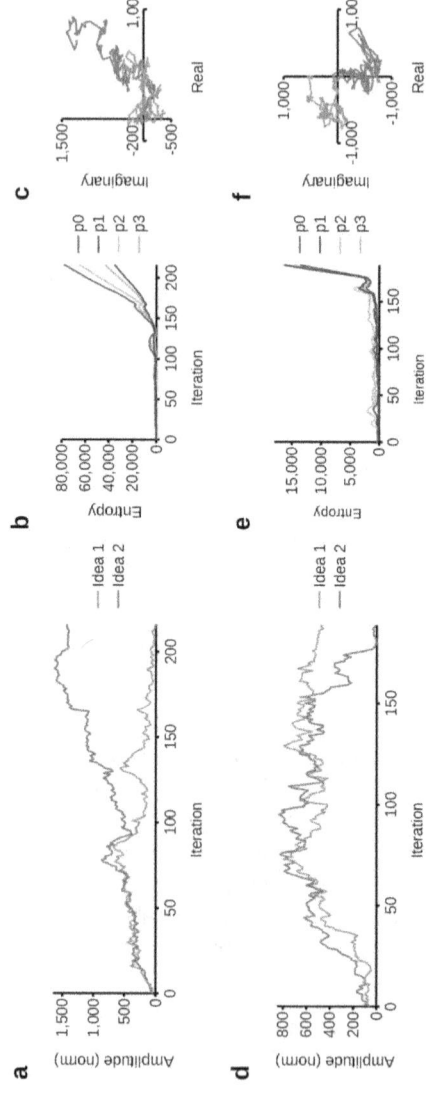

Figure 7: **Competition within a P-swarm.** The diagrams a-c demonstrate the GenI process evolution using a fixed perspective $p = (1, 1)$ under an external environment defined by the observable p_3. Diagrams d-f show another test case under the internal environment defined by the swarm itself. Graphs a/d show the evolution of absolute amplitudes. Charts b/e show the according evolution of entropy for each type according to the swarm member images in $\{p_0, \ldots, p_3\}$. Null pairs for P-swarm do not relate to environment options as is true for E-swarms. Figures c/f display the native paths of each idea in the complex plane.

References

[1] Mukerjee, M. Explaining everything. *Scientific American* **274**, 88–94 (1996).

[2] Stelle, K. Supergravity: Finite after all? *Nature Physics* **3**, 448–450 (2007).

[3] Bern, Z., Dixon, L. J. & Kosower, D. A. Loops, trees and the search for new physics. *Scientific American* **22**, 28–35 (2013).

[4] Hawking, S. & Mlodinow, L. The (elusive) theory of everything. *Scientific American* **22**, 90–93 (2013).

[5] Tuszynski, J. A. The need for a physical basis of cognitive process. *Physics of Life Reviews* **11**, 79–80 (2014).

[6] Hameroff, S. R. Quantum coherence in microtubules: A neural basis for emergent consciousness? *Journal of Consciousness Studies* **1**, 91–118 (1994). Advanced Biotechnology Laboratory, Department of Anesthesiology, University of Arizona Health Sciences Center, Tucson, AZ 85724, USA.

[7] Hameroff, S. & Penrose, R. Consciousness in the universe. *Physics of Life Reviews* **11**, 39–78 (2014).

[8] Tegmark, M. The importance of quantum decoherence in brain processes. *Phys.Rev.* (1999). `quant-ph/9907009v2`.

[9] Hameroff, S. & Penrose, R. Reply to seven commentaries on "consciousness in the universe: Review of the 'orch OR' theory". *Physics of Life Reviews* **11**, 94–100 (2014).

[10] Penrose, R. On gravity's role in quantum state reduction. *General Relativity and Gravitation* **28**, 581–600 (1996).

[11] Penrose, R. On the gravitization of quantum mechanics 1: Quantum state reduction. *Foundations of Physics* **44**, 557–575 (2014).

[12] Kiefer, C. Conceptual problems in quantum gravity and quantum cosmology. *ISRN Mathematical Physics* **2013**, 1 – 17 (2013). 1401.3578v1.

[13] Ellis, J. The superstring: theory of everything, or of nothing? *Nature* **323**, 595–598 (1986).

[14] Tegmark, M. Parallel universes. *Science and Ultimate Reality: From Quantum to Cosmos* (2003). astro-ph/0302131v1.

[15] Tegmark, M. Consciousness as a state of matter. *Chaos, Solitons & Fractals* (2014). 1401.1219v3.

[16] COUZIN, I. D., KRAUSE, J., JAMES, R., RUXTON, G. D. & FRANKS, N. R. Collective memory and spatial sorting in animal groups. *Journal of Theoretical Biology* **218**, 1 – 11 (2002).

[17] Gabora, L. & Kitto, K. Toward a quantum theory of humor. *Frontiers in Physics* **4**, 53 (2017).

[18] Smolin, L. Atoms of space and time. *Scientific American* **290**, 66–75 (2004).

[19] Soklakov, A. N. Occam's razor as a formal basis for a physical theory. *Foundations of Physics Letters* **15**, 107–135 (2002). math-ph/0009007v3.

www.ingramcontent.com/pod-product-compliance
Lightning Source LLC
Chambersburg PA
CBHW050247230526
45470CB00005B/2154